BEACHCOMBERS'
GUIDE TO THE GULF

Tony Woodward

Published with the support
and encouragement of

Published by
Motivate Publishing

PO Box 2331
Dubai, UAE
Tel: (04) 824060
Fax: (04) 824436

PO Box 43072
Abu Dhabi, UAE
Tel: (02) 311666
Fax: (02) 311888

London House
19 Old Court Place
Kensington High Street
London W8 4PL
Tel: (0171) 938 2222
Fax: (0171) 937 7293

Directors:
Obaid Humaid Al Tayer
Ian Fairservice

First published 1994
Reprinted 1996

© 1994 Tony Woodward
  and Motivate Publishing

All rights reserved. No part of this publication
may be reproduced in any material form
(including photocopying or storing in any medium
by electronic means) without the written
permission of the copyright holder. Applications
for the copyright holder's written permission to
reproduce any part of this publication should be
addressed to the publishers. In accordance with
the International Copyright Act 1956 or the UAE
Federal Copyright Law No 40 of 1992, any person
acting in contravention of this copyright will be
liable to criminal prosecution and civil claims
for damages.

ISBN 1 873544 43 X

British Library Cataloguing-in-Publication Data.
A catalogue record for this book is available from
the British Library.

Printed by Rashid Printers & Stationers L.L.C., Ajman.

Contents

INTRODUCTION	4
COLLECTION SITES IN THE GULF	10
WEATHER AND WAVES	14
MOLLUSCS	18
COLLARS, DOLLARS AND SNAKES	28
SANDY BEACHES	32
CREEKS AND MANGROVE SWAMPS	42
FASHT	48
BREAKWATERS AND JETTIES	52
ROCKY BEACHES	56
PRESERVING YOUR COLLECTION	70
POLLUTION	80
MAP	84
REFERENCE MATERIAL	85
ACKNOWLEDGEMENTS	86
INDEX	87

Front cover: The beaches of the Gulf are home to a prolific range of sea creatures.
Back cover: Seahorses may be found amongst grass sea beds, south of Jazirat Al Hamra.
This page: Sandy beach on the east coast of the UAE.

Introduction

Along the banks of the shallow body of water we call the Arabian Gulf, the life of just about every resident is affected by its rich, calm waters in one way or another. Lapping the shores of Iran, Iraq, Kuwait, Saudi Arabia, Qatar, Bahrain, the United Arab Emirates and Oman, it throws up an amazing variety of life along the length of its tideline.

Most evident amongst marine debris are the shells — those beautiful shapes made by a soft-bodied animal, the mollusc, in order to form a hard protective exterior for itself. The different colours and shapes at first seem a little bewildering when trying to sort them into some semblance of order, but they are what initially attracted me to the sea, and with time, my collection was assembled into groups, according to their main characteristics.

● *Conus pennaceus, a beautiful shell made by a lethal stinging mollusc.*

Stop for a minute and look at some of the shells in the 'beachcrest' — the thin brown line which is the mark left by the high tide. If you separate them and then look a little closer while letting your imagination work for a while, you can ask yourself, 'What do the shapes look like?'

A sandy beach near Dibba.

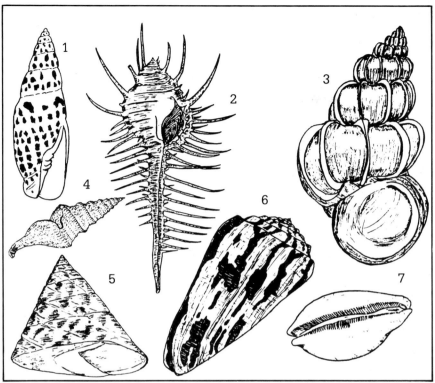

Some of the commonly found shells in the Gulf area: 1. mitre 2. murex 3. wentletrap 4. turrid 5. top 6. cone 7. cowry.

If I tell you that there are turban shells, top shells (as in a child's spinning top), button shells, bubble shells, tusk shells, cone shells and screw shells, hammer oysters, corrugated clams, thorny oysters and sundial shells, can you recognise any of them? They can all be found on most of the shorelines around the Gulf. Once you begin to recognise shapes you can begin to build up your collection and keep it in reasonably good order.

On our journey along the Gulf coast, we will visit a wide variety of natural and artificial habitats, pausing to take a brief look at some

Tusk shells such as this one, Dentalium longitorsum, are frequently washed up on Gulf beaches in their thousands.

of the creatures we can find there. There are far too many species present to allow for inclusion of all of them in this particular book but we are able to give an indication of the common life forms which may be encountered during a stroll along the shore, and to show a few of the more unusual creatures that can be found by a little diligent searching.

A mud creeper, Cerithidea cingulata, lives near the low tide mark.

One of the rarer cowries, Cypraea ziczac, or the zigzag cowry, can be found on the east coast of the Emirates.

One of the pearl-bearing shells, Pinctada margaritifera.

The mollusc that makes the bubble shell can barely retract inside for protection.

Top shells like this one, Euchelus atratus, are very common under rocks and marine debris.

Live Conus tesselatus, are covered with a thick coat called the periostracum.

Epitoneum crassa, one of the wentle-traps, is a rare find.●

There are a number of mitre shells to be found on Gulf beaches, like this Scabricola fissurata.

One of the larger gastropods, Lambis truncata sebae, occurs from the Straits of Hormuz to the Suez Canal.

Mangelia townsendi is a micro-sized turrid shell found in the Gulf. ●

Top shells live in a variety of habitats, like this one which found a home in a piece of sponge. ●

Thorny oysters such as this Spondylis gaederopus, need a solid base to attach themselves to.

Introduction

By the compilation of a checklist, which is highly recommended, you can keep a record of your own finds. If, by chance, you find something not illustrated or described in the book, make a simple sketch, write down the place and date where it was found, and add any other relevant data, such as whether it was alive and if it was high or low tide. At a convenient time take it along to your local natural history group or museum, school or diving club, and they may be able to tell you what it is. Alternatively, send a photograph to Motivate Publishing where it will be passed on to the author.

The Living Seas and *Seashells of the Southern Gulf*, both also published by Motivate, would be useful companion volumes to this guide for enthusiasts wanting to develop further their knowledge of marine life in the Gulf. For those wishing to learn even more about certain groups of animals, a source of reference material has been added in the bibliography.

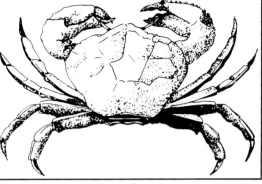

Make a sketch of what you see; it will help others to identify your finds.

Many interesting specimens are washed up on stormy days.

The species Turbo radiatus grazes on algae-covered rocks.

Here the same species extends its foot in an attempt to turn itself over.

Collection Sites in the Gulf

Before going into the Arabian Gulf itself, we'll look at the sites on the Arabian Sea side of the Straits of Hormuz. The high, sharp and jagged mountains of much of Oman's coastline offer little chance to collect except on isolated rocky beaches, and because of the topography access to them is difficult. There are a few muddy creeks, sand flats and mangrove swamps along the shoreline of Oman up to the UAE border which offer better opportunities. Some of the best coral reefs are along Oman's southern coastline where it is possible to find many rare plants and animals.

The high mountains which are such an impressive feature of the region frequently drop straight into the sea. They eventually flatten out as you reach Fujairah in the UAE. It is here, at Khor Kalba, that one of the largest and most interesting of the many creeks in the area is found. Along the banks are mangrove trees that stretch quite a long way inland. There are extensive sandy beaches and as you move further north towards the village of Qidfa, one of the few black sand beaches in the UAE is found.

The mountains once more rise up as you pass the port of Khor Fakkan in Sharjah and approach the fishing village of Dibba which marks the start of the Musandam Peninsula. On the road from Qidfa to Dibba are occasional rocky outcrops which persist as remnants of the former headland, now in the shape of islands. A few interesting and productive beaches occur along this stretch of the coastline, but it is not until you pass through the Straits of Hormuz and enter the Gulf itself, that you find the really long, curving, sandy beaches that are such a favourite place for the family days out.

The beautiful and parasitic desert hyacinth is common in areas of sabkha.

Much of the Gulf's opposing Iranian coastline is similar to that of Oman, with steep mountains and fairly deep water close into shore. On the Arabian side though, the general topography changes quite dramatically as you approach the border between Oman and Ras Al Khaimah in the UAE. The village of Ash Sha'am straddles the border and marks the end of the mountain chain in close proximity to the sea, though it continues to form a backbone along the narrow peninsula. Long sandy beaches and large shallow lagoons can be found and even some low level dunes topped with occasional hardy shrubs. The beautiful parasitic desert hyacinth is common here and it punctuates the sand with its golden flower head.

Just outside the village of Jazirat Al Hamra there is the first of several large muddy lagoons. It runs from the Ras Al Khaimah side of Jazirat Al Hamra to the port (a deepwater channel marks the natural boundary) on the Umm Al Quwain side. The extent is about 10 kilometres in total and it could be described as mangrove swamp, though that brings to mind something

Beaches along the Musandam Peninsula are difficult to reach.

less picturesque than the reality. At dusk this tree-studded sanctuary is alive with the haunting calls of curlews and wading birds, and during the day it bristles with wildlife activity. Insects of many sorts are present, particularly cicadas and dragonflies. Invertebrates such as crabs and worms abound, and more terrestrial insects such as moths and butterflies, grubs, wasps and ants are resident. There are tree rats and gerbils, not to mention snakes. A little further south of the village are the first large grass beds which are home to some unusual marine animals such as sea horses and sea hares.

The shallow bays continue until you reach Umm Al Quwain. Here an experimental fish station has been started, testament to the calm nature of the sheltered bays. The sea is quite shallow and the coastline is indented with several large creeks which are capable of taking quite considerable-sized craft inland, though periodic dredging is necessary to keep them from silting up.

Ajman and Sharjah also have creeks which can accommodate a wide variety of large industrial vessels as well as smaller dhows. These beautiful timber ships still ply the time-worn routes to the Indian sub-continent, East Africa and Iran as well as to other parts of the Gulf.

Once you reach Dubai there are more opportunities for collecting in artificial habitats than in pristine natural ones. This well-established area is fast losing its best beaches apart from short stretches at Jumeirah and Umm Suqeim. Perhaps one day the government will create zones the other side of Jebel Ali port, some 40 kilometres from the town, for the casual stroller to enjoy.

The long Abu Dhabi coastline offers more chance to collect on sandy and muddy beaches than on rocky ones. It is the first really extensive area

Collection Sites in the Gulf

where tracts of fasht can be found. This limestone rubble will be discussed later in the book.

At the border between Qatar and the UAE is a deepwater inlet known as the 'Inland Sea' or Khor Al Odeid. There are steep dunes on the Qatari side with an average water depth of from 10 to 16 metres at the edge of the dunes to 26 metres at the mouth of the inlet.

As you pass the port of Umm Said there are some rocky reefs and sand-bars which are shallow enough to walk out to at low tide and so are ideal for beachcombers.

North of Qatar's capital city, Doha, there are some long stretches of fasht and large sand-bars where numerous species of sealife can be collected during the course of a day. On past Al Khor the surrounding flat stony desert is punctuated with outcrops of rock, particularly at the coastline near the village of Fuwairat. It is one of few such features in the country. Around the tip of the peninsula you reach very shallow water where the average depth is just five metres but it has some interesting raised fossil beaches near Bir Zekrit, especially at Ras Abruk. The shallow bay of the Gulf of Salwa forms the border between Qatar and Saudi Arabia. It has such high salinity that it might be assumed few animals could survive. Yet at the nearby Saudi Arabian terminal of Ras Tanura, experts at the ARAMCO headquarters have found that there is a richness almost unsurpassed in the number of species so far recorded. The Gulf species count of over 1,000 in respect of molluscs alone, exceeds that of any other similarly sized body of water (including the Mediterranean) in the world. Overall, the marine species, though not as spectacular or as large as in the neighbouring Indian ocean, are more diverse and certainly more numerous.

The 'lost river' is an inlet near Dhakira, in Qatar.

The low-lying Bahrain archipelago offers a considerable number of habitats to explore, with the notable exception of rocky beaches. The best shell beaches occur in the south at Ras Al Barr.

Most of the coastline of Saudi Arabia towards Iraq is little different in that there are shallow lagoons and flat sandy beaches with occasional mangrove swamps.

Kuwait and Saudi Arabia both have extensive mud flats and offshore islands where residents of the nearby towns can pursue their hobby.

Collection Sites in the Gulf

Finally, at the headwaters of the Gulf, the Shatt Al Arab waterway is the only freshwater inflow of note. Though its existence is of great commercial and military importance, the effect it has on the salinity level of the Gulf is quite minimal. Obviously there are still areas around Kuwait and southern Iraq in which it is difficult to collect due to security reasons. In time these areas

A shell beach is best investigated at low tide.

may become available again and their richness rediscovered. It is best to consult the local authorities and obtain written permission in advance if you intend visiting, so as to avoid disappointment.

Children spend hours looking at the variety of shapes in beach debris.

2 Weather and Waves

The Arabian Gulf is shallow; in fact, it rarely exceeds 30 metres in depth except at its open sea entrance, where, at the Straits of Hormuz, the depth is 90 metres plus. Strong currents cause the water to rush past the jagged, towering cliffs when the tide changes; just like water passing from the narrow mouth of a bottle. Unfortunately there are few beaches in this area which are accessible to the collector, but I have made a number of dhow trips into the Straits of Hormuz and along the coast of the Musandam Peninsula. It is a quiet and eerie place where the only dwellings consist of almost hidden, rather primitive stone-built villages. It is as if the inhabitants dare not disturb the omnipresent air of tranquillity. The occasional scream of a sea bird or the crash of a large predatory fish as it leaps after its nimble prey are all that pierce the silence.

The tidal inflow brings with it a rich planktonic soup from the Arabian Sea and the Indian Ocean to nourish the sealife that is found on the subtidal rocks just inside the Gulf. As the water rushes along the mountainous Iranian and Ras Al Khaimah coastlines it gouges out debris and turns it into silt to be deposited on the coastline further north.

Once the incoming tide enters the southern end of the Gulf, it starts to decrease in speed. It then swirls in a series of anticlockwise spirals, snatching up minute fragments of rock and debris from the deeper Iranian side, and depositing them on the more shallow Arabian side of the Gulf, hence the large sandy beaches that we enjoy so much. There are times, however, when a strong wind can overcome the speed of the flow, causing a surface current to run in the opposite direction from the main tidal stream. At these times the weathermen must despair, and we collectors too — for the prediction of tides will be sadly awry. Luckily, however, the tidal range (the difference between high and low tide) in the Gulf is slight, so we can collect on most days and at most times.

A summer storm over the mountains.

Stone-built villages cling to the rocks of the Musandam.

If you do consult tide tables (the daily high and low tides are given in most Gulf newspapers) you will see that there are certain days during each month when the highest high tides and the lowest low tides occur. These 'spring tides' are worth considering because at low tide they create an unusually large area of exposed beach for the collector to survey. However, you are well-advised to double-check your equipment at these times, especially if you are in a four-wheel-drive vehicle and going to a remote place. Once in Qatar, a few years ago, I failed to do this — with disastrous consequences.

I used a friend's vehicle when investigating a sand-bar which was only exposed at spring tides and where I had made some really interesting finds in the past. I got carried away with the collecting and ignored the warning signs of the in-coming tide cutting me off. In a desperate attempt to get off the sand-bar, I became bogged down. By the time I returned with help, the lowest low tide had become the highest high tide and all that was visible of the vehicle was its roof! Since then I have remained rather cautious of the places I explore when there is a spring tide, and make sure that the vehicle is properly equipped with a tow rope, some boards to put underneath the wheels, and a jack.

Rough days, like this one at Sharm in Fujairah, are ideal for beachcombing.

Weather and Waves

On days when there is a strong wind blowing, not only is the tide difficult to predict but we hardly feel like going out to brave the swirling sand in our faces. Yet after a stormy day there will certainly be a few unusual finds to be made. Thanks to an angry sea, weed and sponges are snatched from the seabed and flung far up onto the beach to become part of what is often called flotsam and jetsam. The shamals are very strong winds that blow over the Gulf in summer, and though the term is variously applied to winds which originate in the north (not necessarily blowing from that direction), or to winds of over force five in strength, their effect is the same — joy to some and misery to others. The 40-day shamal, known as Barih Al Kabir, occurs roughly between 6th June and 16th July. It is a time when most water sports drop to a minimum and therefore it is an ideal time to get out those beach collecting bags. The effect of this prolonged wind is greater in the northern and central Gulf region — Kuwait, Iran, Iraq, Saudi Arabia, and to some extent Bahrain and Qatar — than in the south around the UAE and Oman. However, for those willing to go out during the calmest time of the day — early morning — there are some pleasant surprises in store at this otherwise rather hostile weather period.

One year, in the early '80s, the Barih Al Kabir actually did last for about 40 days. At that time I was in Doha, Qatar. It was a misery because the water turbulence caused poor visibility on most days, preventing us from either diving or from venturing too far offshore. The moaning wind caused a similar reaction in most of the residents but some pleasure came from wandering along the sandy beaches near Dhakira and the Al Khor coastline. When the wind dropped sufficiently for the ruffled water of the tidepools to become calm, it revealed the tiny tracks of sand-dwelling animals and allowed time for me to observe the numerous creatures that I had previously neglected.

This offshore rock, close to the Sandy Beach Motel in Fujairah, is one of the most interesting sites for beachcombers to explore along the east coast.

Sand-bars are formed by the action of the waves and are ideal homes for bivalves.

3 Molluscs

THE FIVE MAJOR CLASSES OF MOLLUSC FOUND IN THE GULF

The chart below shows the scientific divisions of the animal kingdom.

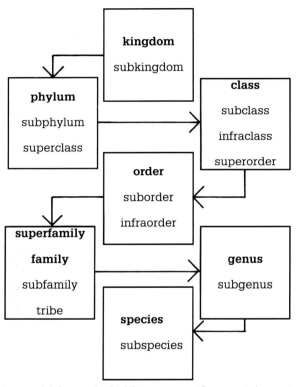

The divisions which are in bold type, are the most important and the ones which are most frequently referred to in books.

As you will see, after the kingdom is a division called a phylum (plural phyla), and in the animal kingdom there are some 30 phyla known today. The largest in terms of number of known species is the phylum Arthropoda (to which spiders, crabs and insects belong), and the second largest is the phylum Mollusca.

There are approximately 160,000 species in the phylum Mollusca and this number is being added to and amended constantly, as material becomes available to experts in the field of malacology (the study of molluscs).

The species are grouped into six classes within this phylum, and only one of the classes (Monoplacophora) is not represented in the Gulf. It is very obscure and comprises only a few living species found in water over 3,000 metres deep.

This chapter is a guide to the five classes of mollusc that do exist in our area.

The six known classes of mollusc.

1
2
3
4

1. Monoplacaphora
2. Amphineura (Chitons)
3. Scaphopoda (Scaphopods)
4. Gastropoda (Gastropods)●
5. Bivalvia (Bivalves)●
6. Cephalopoda (Cephalopods)

5

6

Molluscs

GASTROPODS (SNAILS AND SLUGS)

As this class is the largest, it is perhaps the easiest to recognise and we will deal with it first.

Most representatives of the gastropod class have a single shell and hence they are sometimes referred to as univalves. The most primitive gastropods are the limpets and abalone families. They progress in terms of development through the winkles, conchs, cowries and moon snails until we arrive at the more advanced families of the cones, mitres and auger shells. At the upper end of the shelled gastropod class, there are families where the shell plays a less important role and the animal has started to evolve to the point where it no longer needs the protection of a thick outer surface. Such is the case with the bubble shells and some of the air-breathing snails which are not only starting to lose their need for a shell but in some cases can live by breathing air, rather than siphoning water for respiration.

The most advanced gastropods either have a fragile internal shell, such as is found in the sea hare, or have no shell at all, as is the case with the sea slug and its land-based relative, the garden slug. This subgroup of gastropods is called the opisthobranch subclass and they have developed a complex, alternative method of protecting themselves. Some of the brilliantly coloured sea slugs can reuse the stings of other invertebrates on which they feed by storing them up in their bodies. Others are able to secrete acid to deter attacks by would-be predators. Truly, they have entered the chemical and biological warfare era.

Nudibranchs or sea slugs are snails without a shell.

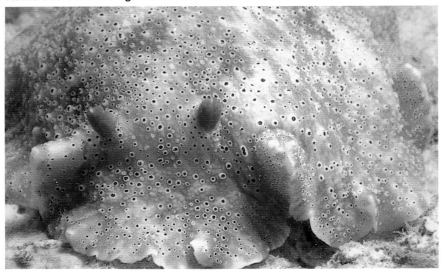

Thrush cowries, or the species Cypraea turdus, are the commonest of the Gulf cowries.

Of all the gastropods, the beautiful, shiny, ovate shells of the cowry family display an unparalleled appeal because of their exquisite patterns. The word cowry, or cowrie, apparently comes from the Hindi language and certainly they were used as a form of currency in many parts of the Indian subcontinent, as well as in south-east Asia. Perhaps many sacks of the 'gold-ringed cowry' or the aptly named 'money cowry', changed hands in days gone-by to enable trade to be carried out in its earliest pre-hard-currency form. Both species occur only rarely around the Arabian Peninsula, though they are often found as an adornment on old horse bridles, necklaces, jewellery and clothing in Yemen and Oman. Perhaps, generations ago, they were the object of trade between the Arabian countries and India or Africa, where they are abundant. They were probably transported in dhows and traded for coffee and spices.

A pair of Carnelian cowries, Cypraea carneola, completing their egg-laying.

There are several other species of cowry shell to be found in the Gulf and by far the most common is the 'thrush cowry', Cypraea turdus. This bluish-grey shell has a white base and dark spots over the back and sides. This makes it quite distinctive, yet when the live animal covers its shell with its mantle, it becomes almost impossible to see against a background of weed

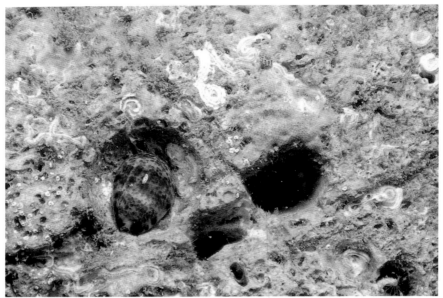
Cypraea felina fabula or the fabulous kitten cowry, found only in Oman and the UAE. ●

and branching bryazoa. It is thought that, as a means of defending itself, the cowry may suddenly reveal its bright contrasting shell to scare off a would-be attacker, such as a fish or crab. However, this is only supposition as there are some cowries that have an almost transparent mantle that does little to hide the shell underneath. Some species can be seen browsing around in the daytime, making no effort to conceal their shell pattern, so maybe they have developed another method of protection which remains to be discovered. Whatever the reason, it is certain to fascinate you if you come across one in the live state.

Apart from the thrush cowry, which is found throughout the Gulf and along the shores of the Gulf of Oman, there is another species, the 'freckled cowry', Cypraea lentiginosa, which, if not indigenous to the Arabian Gulf, certainly has its main centre of population here. Much smaller and far less common than the thrush cowry, this cowry lays its orange egg mass in April and May underneath stones in the shallow waters of the Gulf. It has been recorded in Kuwait, Saudi Arabia, Qatar and the UAE to my certain knowledge, though, strangely, not along the eastern coast of the UAE. It reappears in Pakistan and the western coast of India but in a slightly different form.

Another of the cowries which is definitely confined to a narrow population centre is the 'fabulous kitten cowry', Cypraea felina fabula. Cypraea felina felina, the 'kitten cowry', is a common Indo-Pacific inhabitant but C. felina fabula is a more ovate and thickened form that is only found along the east coast of the UAE and into Oman. It is quite distinct and seems to have grown into this shape because of its isolation from the other more common subspecies.

BIVALVES (MUSSELS, OYSTERS, SCALLOPS AND CLAMS)

These are the familiar clams and oysters and they are evident in almost every zone along the seashore. Some are of considerable commercial value. The famous pearl oyster is one that was harvested in large quantities in times gone-by and is easily recognised as a representative of this class. All bivalves have shells or valves which are in two pieces which may or may not be equal in size. They are hinged along one edge by a ligament and open and close by means of one or two strong muscles called the adductors. After the animal dies, a scar is left where the muscle was attached to the inside of the valves and its form is different for most species. Together with the general shape and colour of the bivalve, as well as another scar which forms a line where the mantle tissue was formerly attached, the muscle scars are a key part of the identification process. An additional feature that helps to identify the many species, is the shape of the teeth that form where the two valves lock into each other. For the most part they lack eyes though there are some species which use light-sensitive cells to tell the animal when it is time to close up.

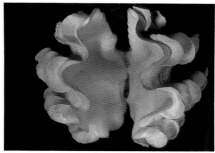
Fluted clams (Tridacna squamosa) are amongst the most attractive bivalves.

Venus clams, from the largest bivalve family.

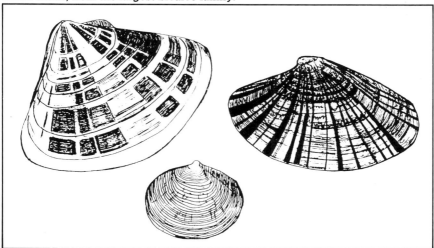

Lima fragilis, a flame scallop, is one of the few bivalves able to swim for short distances by clapping their shells together.

Another species of flame scallop, Lima sowerby, is sessile, attaching itself to the underside of rocks.

Many families are sessile, that is, they live out their lives in one spot usually attached to the rocks by threads of fibrous material called a byssus. This allows the animal some freedom to change positions without being washed away by the tide. However, there are some, for example cockles, that can crawl about, using a well-developed foot. At least two families, the scallops and the file shells, are able to produce short bursts of swimming to escape predators.

SCAPHOPODS OR TUSK SHELLS

Although this is a small class, it is well-represented in the Gulf and the surrounding areas. You will be sure to find a number of these curious shells on most of our beaches, and they really do look like miniature elephant tusks.

In the Gulf, scaphopods can be found in water only two metres deep but in the Mediterranean, for example, they are recorded in depths of over 200 metres.

There are three species of scaphopod known in Gulf waters. These Dentalium longitorsum came from Qatar.

They are covered over by sand and feed on tiny creatures called diatoms which share this habitat. One species that is very common in the area has eight sides and therefore has earned the name Dentalium octangulatum.

POLYPLACOPHORES OR CHITONS

The Gulf region has several species of chitons, some of which are only familiar to readers of technical publications. They look rather primitive and can even be mistaken for fossils when seen pressed onto the rocks. They are almost impossible to dislodge without causing some damage and if you do manage to get one away from its

The largest of the primitive chitons found in the Gulf, Aconthopleura vaillantii, is found on exposed rocks.

home, it immediately curls up to protect the soft foot underneath with its segmented shell.

We will discover more about these strange animals later in the book (see chapter 8) but one thing worth remembering is that they are more active at night. If you go down to the shoreline after dark you may see them making their way slowly across the surface of the rock, grazing on algae. Before dawn they return to where they started from, in order to rest as the day becomes too warm for movement.

CEPHALOPODS (SQUID, OCTOPUS AND CUTTLEFISH)

This is the most advanced class of mollusc and one which has a number of intelligent families that can swim quite fast in order to catch their prey. Most representatives do not possess a shell, though the argonauts and the cuttlefish are exceptions. There are members of each of the main families present in the Gulf and the surrounding areas, but the only ones that leave evidence of their presence on the beach are the cuttlefish and two species of argonauts known as 'nautilus shells'.

An octopus, most intelligent of all molluscs, if not all invertebrates.

The argonaut has a shell which creates much interest, and its life cycle is fascinating. The female, which is quite large, with an egg case measuring 50 to 80 millimetres, makes a basket under her body to keep the developing offspring in until they are ready to leave and fend for themselves. The male who, by comparison, is minute (5 to 10 millimetres), clings to the egg basket until the female is ready to receive him. He then fertilises the eggs and his sole task in life is done. Once the eggs hatch and the young grow up sufficiently to venture out of the safety of their home, the female jettisons the shell or basket, and along with it goes the hapless male. It is this egg case that can sometimes be found on our beaches during the early part of the year.

Squid, such as this Loligo sp, are of commercial importance.

Sepiella immanis, the common cuttlefish, is unafraid of humans.

Collars, Dollars and Snakes

I was sitting on a beach recently near the borders of the UAE and Oman at the village of Al Rams, when a man approached me with what looked like a piece of string on the end of a stick. As he came closer I realised that, in fact, he had found a washed-up sea snake and thinking it was dead, was about to show it to his children. When I told him that the creature was probably not dead as they often get washed ashore during storms, and that if he did not want his children to receive serious bites he should let it go back to sea again, he looked more closely but with obvious disbelief at this inanimate reptile. Then suddenly it moved a fraction, which made him decide to toss it promptly back from whence it came.

The Gulf appears to be a site of major importance for sea snakes, and there have been occasions when I have seen literally thousands of them congregate together offshore. At these times the sea looks like a bowl of spaghetti. These rather unloved air-breathing creatures are often sighted as they surface to breathe, in the shallow areas near the beach. During moderate to rough seas they are frequently washed up onto the beach and there they wait until the high tide carries them off again, if they survive that is. They are absolutely helpless on land even though they evolved from land-based reptiles (phylum Vertebrata, class Reptile) and it can be a common mistake to think that they are dead. They also possess a venom which is many times more powerful in toxicity than even the most deadly of land or elapid snakes. Care should obviously be taken when they are encountered and if you have to move them for the safety of people on the beach, use a stick and manoeuvre them carefully back into the water.

Something less deadly but still a fascinating object, and one which can be taken home and cleaned easily, is the sand dollar. These flattened discs are generally found on sandy beaches though they have been known to congregate around the edges of reefs and submerged islands too. They are close relatives of the sea urchin and have a most impressive flower-shaped pattern on their backs. The live sand dollar is able to move about on minute legs and it makes quite a wide track when submerged just below the surface of the sand. There are five known species here in the Gulf, though there may be more waiting to be discovered. They belong to the Echinoidea class of the phylum Echinodermata.

The Gulf appears to be of major importance for sea snakes.

Frances Dipper

The spines of the Diadema urchin can inflict painful wounds.

If you find one that is not long dead or still alive then take a look under the disc and you will see one large central hole that looks as though it is encircled with teeth. This is referred to by scientists as Aristotle's lantern, after the famous scholar and one of the world's first naturalists. It is one of the most fascinating mechanisms in the animal kingdom. Basically there are five teeth with a complex structure of 40 calcareous pieces and a number of muscles. It allows the sand dollar to grasp and chew its food by moving the teeth up and down and toward and away from each other.

Unfortunately, the chances are that any you find will have been dead for a while and will already be partially bleached out. They can be

It is hard to believe that sand dollars are 'flattened' urchins. ●

rather fragile in this condition so care is needed if you are to get them home without breaking them. The cleaning process is simply to immerse the dead dollar in household bleach about 70 per cent strength, for an hour or two at the most, and the whole thing will be gleaming white. After drying out it will become ready to take its place in your collection.

Collars, Dollars and Snakes

Collars are another interesting find. Very often you will see an odd structure washed ashore that looks like a sand collar or a necklace. It is in fact the egg case of the moon snail — a predator which will be discussed more fully later (see chapter 5) and which is also known as the necklace shell. It makes these elaborate structures by mixing mud and sand from the seabed, together with mucus that the animal secretes, and then it deposits its eggs amongst them. Once the baby moon snails are hatched, the egg case loses its use and slowly drifts towards the shore. It will break up unless it is collected carefully.

There are many species of moon snail in the Gulf and its surrounding waters, and quite which makes which type of collar is yet to be determined. In more than 10 years of underwater observation I have yet to come across one being made, so perhaps they are created when the snail is under the surface of the sand.

A stealthy predator, Natica alapapilionis — *this moon snail drills holes in the shells of other gastropods.*

The sand collar of Polinices tumidus, photographed in Khor Khasab.●

A sand dollar clearly showing the distinctive pattern.

5 Sandy Beaches

Collecting on a sandy beach may at first look a daunting prospect, with not much to see, but let me give you an insight into what lives there and how to find it.

Ghost crabs are responsible for the small pyramids of sand seen on beaches.

First look at the beach. You will notice that there is a thin brown line that meanders along it. This is called the beachcrest. As it is the mark of the high tide it would seem a good place to start our search. Sit down and take your time when examining the dried weed which makes up the largest part of the crest. There are sure to be some shells and often the weed itself may prove attractive. Here and there small pieces of sponge and the carapace of a pretty little crab can be uncovered. There are places, such as Umm Al Quwain in the UAE, where the beachcrest comprises a mass of small pearl oysters (Pinctada radiata) which children can arrange into a collage with other bits of debris on days when they are housebound. You may well find a clump of pink or grey barnacles and these too can be used to good effect with driftwood in floral displays. Collecting from the beachcrest does no harm to the ecology, indeed the worst offence you could be guilty of is depriving a hermit crab of a possible new residence.

Cross-section of a typical sandy beach.

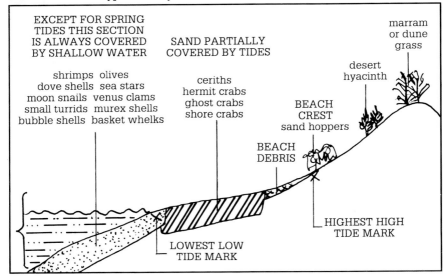

A sandy beach at Dibba.

The best places to look for shells are in the sandy areas near shallow coral reefs, and along those beaches where sand gets mixed with mud. This provides a rich feeding ground for many molluscs and, in turn, other life-forms.

Around Dibba, on the east coast of the UAE, there is a long curving beach where I first found specimens of a gastropod called the wentletrap or spiral staircase shell. Wentletraps are so unusual that in days gone-by collectors paid fabulous sums of money for the shell now known as Epitoneum scalare or the 'precious wentletrap' which is found in the Pacific Ocean. The Kaiser Franz I Stephan (husband of Maria Theresa), is reputed to have parted with 4,000 guilders for a fine example — and this was in 1750! There are several species found in the Gulf and its surrounding waters. One, Epitoneum pallasi, is similar to the 'precious wentletrap' but, unfortunately, it is only rarely recorded.

Wentletraps are strange creatures in their true habitat. They live on anemones and inside other invertebrates, sucking out the juices of their unwitting host in a parasitic manner. Many live so deeply embedded that they cannot be found except by removing the complete animal that houses them and cutting it open. That is not a practice to be condoned, unless undertaking serious scientific study, so we have to wait until either the host or its parasitic wentletrap dies and then gets washed ashore.

A rare wentletrap, Epitoneum crassa, seen here laying eggs in a sandy lagoon at Ras Al Khaimah.

Peacock's tail weed often forms a major part of the beachcrest on the east coast of the UAE.

The ubiquitous pearl shell, Pinctada margaritifera, like this one at Sur in Oman, is often washed up during storms.

Sandy Beaches

One sharp and spiny resident found on most sandy beaches is a shell that could make itself known in a somewhat unpleasant manner, should you inadvertently step on it. I found one by this rather unconventional method many years ago on an island called Masirah off the Omani coast and have since discovered it to be a gastropod called a spiny murex. There is an even spinier character in this family called a Venus comb (Murex pecten) which can be found in Sri Lanka and generally throughout the Indian Ocean and the Pacific region, though not here in the Gulf. There are two established species of spiny murex in the Gulf, Murex scolopax and Murex aduncospinosis. The most common is Murex scolopax but they are rarely found with all of their spines intact. They live in the mud and come out to hunt at night. When they find a clam they force its valves open to feed on the unfortunate bivalve within. Perhaps this is one more reason why there are so many bivalve shells washed up on the beach.

Found on a Kuwaiti beach, this common spiny murex, Murex scolopax, represents one of three species in the Gulf.

It is easy to see why this shell is called the Venus comb (Murex pecten). It is not found in the Gulf.

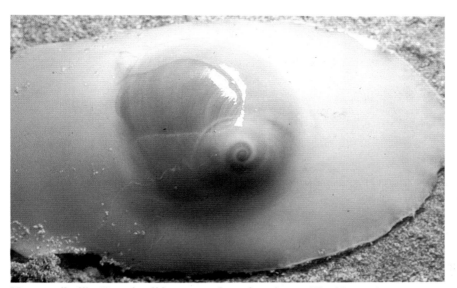

Moon snails, like Neverita didyma, are also referred to as 'Shark's eyes' in some publications.

The murex is not the only predator that lurks in the shallow sandy water though. There is one gastropod that has developed a rather special way of getting its fill. The moon snail is a bright and shiny shell often with an attractive pattern, but with something of a Jekyll-and-Hyde character. I watched one moving over the sand one day at the Inland Sea or Khor Al Odeid in Qatar. The animal had the grace one would expect from such a beautiful shell. The creamy coloured animal simply seemed to glide along over the sand. Yet it was able to capture bivalves and tiny gastropods by smothering them with its large fleshy foot. Once it traps the prey, the moon snail is thought to emit an acid secretion from a special gland located in its foot. This reacts on the calcuim of the shell and makes the small neat hole we often see in the shells on our beaches. Once it has made the hole, it is then able to suck out the creature from inside. Just take a look at the shells on the beach and you will be able to pick out some with small holes in them. Some members of the murex family are able to drill holes too in similar ways to the moon snail.

Not everything you can find on the beach is in the form of dead shells and weed. Crawling around in the beachcrest are some very lively little creatures. They often

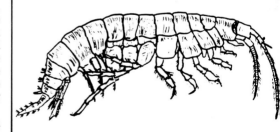

Sand hoppers or sand fleas live amongst the beachcrest and cause skin irritations when they bite humans.

make their presence felt if you happen to get amongst them, as their name of sand hopper or sand flea suggests. There are quite a few hermit crabs hiding under the weed too. They protect themselves in this manner from the heat of the day but at night become very active as they search for detritus (rotting material such as dead crabs, fish and sea grass) along the beach.

Blue swimming crabs shed their shells frequently. They are washed up and can be found amongst the beachcrest.

Tangled in some of the weed you are quite likely to spot the carapace of a washed-up crab. In the Gulf and the surrounding waters there are many colourful crabs and one in particular, the blue swimming crab, is of some, although limited commercial importance.

It is time now to stretch our legs a bit and move down the beach a little closer to the water's edge. The area will often still be damp from the receding tide. While the tide is out, there is an opportunity to find some of the tiny creatures that dig into the sand to take refuge and keep cool until the tide comes back in. Tracks are the tell-tale marks that give away their hiding places and so we will have a look at a few of them now.

The hermit crabs that I mentioned before have quite distinctive tracks which look as if a mini-tank has passed by. A thin line drawn in the sand usually identifies that a cerith shell has come to a halt, but at which end of

Weed is continuously washed ashore.

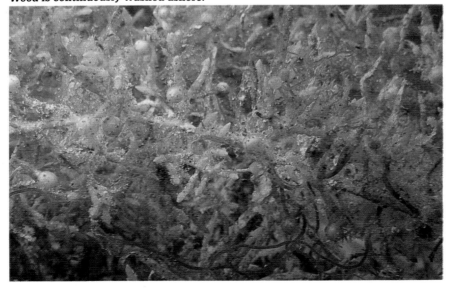

Cerithidea cingulata making tracks in the sand.

the track? If you check both ends, you are sure to find it. A thicker track, in a fairly straight line, often means a large olive shell has come to rest. These glossy and solid-shelled gastropods have a highly variable colour pattern and a passion for small bivalves which they trap with their large slimy foot. When it comes to its breeding season, the olive uses the bivalves for something else other than its main source of food. This extraordinary mollusc catches a couple of bivalves and then attaches the start of its egg string to them. The olive then releases the bivalves, which not able to believe their luck, quickly bury themselves into the sand. With them goes the end of the egg string. The olive then stretches out the string along the seabed and collects more hapless bivalves which it lets go, after securing the egg string to their shells. The bivalves repeat their digging-in trick and so hold the other end of the string firmly in place. This unusual use of what would normally be considered a major part of this predator's diet has not been observed in any other family to my knowledge. I am indebted to Kathy Smythe for her observations of this behaviour which she described in one of her many papers on the molluscan fauna of the Gulf.

Check both ends of tracks made in wet sand and you may well find a beautiful olive shell such as the Oliva bulbosa.

Sandy Beaches

The species of olive found most commonly in the Gulf and its surrounding area is Oliva bulbosa, 'bulbous olive', but you may well find one of the other species if you look carefully. One, Ancilla castenea, comes in three colour forms (chocolate brown, golden and white) and often finds its way into very shallow water where it can hide in the wet sand at low tide. An even smaller species is Ancilla ovalis which also lives in the wet sand and makes a very thin track. It is present on most, if not all, sandy beaches in the Gulf but unlike the other olives, this one has a very thin operculum or trap-door. Present in many gastropods, this trap-door is composed of either a horny material or else is thin and papery. It is designed to seal the aperture of the shell and so protect the animal against attack by predators but in this case the olive has evolved to become a predator itself and rendered the operculum almost redundant, hence its very thin, almost non-existent state.

This tiny olive shell, Ancilla ovalis, can be found in a variety of different colours.

Sea cucumbers are found partly buried during the daytime and eat huge amounts of sand and mud, sifting out tiny organisms.

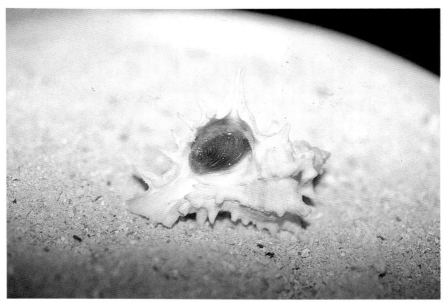

Muricanthus kusterianus, the most common murex in the Gulf, lives on a diet of bivalves in muddy and sandy habitats.

There are plenty of top shells, another gastropod, scattered around, often living on isolated clumps of weed or under an occasional stone or rock. Maybe you will find a murex or two which at the time of an extremely low tide could be alive. One of the bivalves that is a familiar sight for collectors all along the Gulf coast, is also known to many motorists. Take a look at Ruschenbergs scallop — it is sure to be there — and you can see the inspiration for the Shell oil logo. They are mostly dark red or brown but some albino examples occur on the eastern coast of the UAE.

Another bivalve capable of swimming by clapping its valves together is the Ruschenbergs scallop.

Perhaps the most easily recognisable shells to be found, in the Gulf area at least, are the thorny oysters or Spondylis gaederopus. They retain their glorious colours even after death, and you can collect them in hues that range from white to blood red, pale purple and orange, or in a combination of colours. They grow in very dense numbers on wrecks and rocks, and have the habit of squeezing their neighbours out of their living space. The evicted are washed ashore to form a dazzling addition to the tideline.

6 Creeks and Mangrove Swamps

When, in the 60s, I first saw trees growing in the sea as I wandered along the beaches in Bahrain near the village of Bu Ashira, I must admit to being a bit bewildered as to how they could tolerate the salt water and yet look so healthy. I later discovered an explanation for their unconventional life cycle. Firstly, the roots are called pneumataphores because they are air-breathing. They support the tree which periodically flowers and produces a fruit, shaped like a torpedo. When this embryonic fruit reaches a few centimetres in length, it drops off and hopefully sticks into the mud. If it does, then a new tree will start to form. At first the leaves grow from the top and then aerial roots gradually form at the mid point. The roots eventually reach the mud and then the main root, or taproot, dies and the tree is supported by these aerial roots. The whole thing looks a bit like an umbrella skeleton at the base of a fully mature tree.

Trees that grow in the sea — mangroves.

Shallow mangrove swamps in Oman are home to a host of marine creatures and many birds.

Dubai Creek, a busy commercial waterway, with an array of interesting marine life.

Creeping around the roots of mangrove trees is not to everyone's liking but for those who do venture in, there are plenty of things to see. The bird life can be very varied but not because of the trees themselves, as the lush leaves are very acrid and full of tannin. The reason is, as you might expect, the wealth of life that is attracted to the tree for protection. Tiny creatures called fiddler crabs use their pincers — one huge and one small — to sift through the mud and leafy debris under the trees. The larger blue swimming crab is also found here; partly buried in mud but ready to spring out with its spiny pincers opened in a defensive manner if you get too close. Plenty of bivalves bury themselves in the mud too, so wading birds are a common sight in the swamp as they seek them out. One bivalve which was found more commonly, when collecting was permitted in the swamps near the Shatt Al Arab waterway in the northern Gulf, is the window pane oyster, Placuna placenta. It gets its common name from the fact that it was used as a glass substitute in both Asia and China in former times.

A mass of cerith shells of the species Cerithium scabridum, forming a carpet in one of the Emirates' many creeks.

With patient observation, a considerable number of animals and birds can be recorded in the swamps. Birds such as the reef heron, plover, curlew or even a rare kingfisher may eventually make themselves known. There are plenty of cerith shells and mud creepers lying around and one of them,

Creeks and Mangrove Swamps

Cerithidea cingulata, is often so prolific that it may form a living carpet on the mud. It was here in the creeks that the ancient inhabitants of the Ras Al Khaimah and Fujairah coastlines collected large numbers of an edible cerith-like shell of the Potamididae family called Terebralia palustris. Piles of these shells, long-dead, can still be found in Ras Al Khaimah and Fujairah today and this brown, horn-shaped snail is also favoured by hermit crabs. Terebralia palustris is thought to be extinct, as far as I know, in the Gulf itself but has been reported living in swamps further south in Oman.

Little dove shells with their distinctive zigzag patterns move about in search of detritus to feed on. They share the territory with another gastropod, the basket whelk. By waving their siphons around like miniature elephants holding up their trunks, these basket whelks constantly test the water around themselves to see if a chance meal can be found.

A tiny dove shell, Mitrella blanda, common in creeks and shallow swamps.

This hermit crab is using the shell of Strombus decorus persicus.

Cluster winks or false winkle shells, Planaxis sulcatus, festoon the pilings of a creekside wharf.

Fishing dhows moored in Ajman Creek.

Creeks and Mangrove Swamps

The roots and stems of the mangrove also provide an opportunity for attachment and here, without too much effort, can be found barnacles, oysters and clusters of planaxis shells. They all seem to be able to survive considerable periods of time without needing the tide to wash over them. The barnacles, you may well be surprised to discover, are more closely related to crabs than they are to molluscs. They are members of the crustacea that have evolved by making a calcareous shell around their legs. They live out their lives waving distinctive tiny tentacles called 'setae' in shallow water during high tide in order to catch the microscopic organisms that float past. When the tide eventually oozes out, they curl up, safe in the knowledge that with the combined protection of their thick shell and inaccessible habitat, they can rarely be attacked.

There is one resident of this type of habitat that is threatened with something more sinister than a casual predator — extinction. Small colonies of tiny gastropod molluscs that resemble winkles at first glance, may reveal themselves to be the air breathing snail known scientifically as Salinator fragilis. They can still be found in a few places in the UAE but unfortunately the colonies are getting smaller both here and worldwide, though for what reason it is not certain. It is hoped that people who collect in the mangroves and creeks of the Gulf will be wise enough to leave alone the remaining live animals and take only dead samples.

Sudden proliferations of the bubble shell, **Hydatina physis***, occur in creeks every few years, though the reason is unknown.*

Creeks and Mangrove Swamps

Often washed up along the edges of the creek and in the muddy mangroves, is a shell known to many who may never have seen the sea. It is none other than the cuttlefish shell, or 'bone' as it is frequently called. This lightweight white piece of flotsam can easily be snapped in two. Should you do this you will see the layers that the cuttlefish secretes from special glands in its mantle. This mucus secretion immediately crystallises and builds up into layers within the oval shape of the shell which many may only know from pet shops where it is sold for caged birds.

It was along a beach near to Khor Kalba, perhaps one of the most unspoilt natural creeks in the UAE, that I found an odd-shaped bone, almost like a star, and quite large. Later I found the remains of a dead green turtle. Often these lumbering giants are discovered after they die from the effort of dragging themselves ashore to lay eggs. The shell or carapace of the turtle has four interlocking bones encircled by a gristly girdle and then covered over with the more familiar mottled brown 'turtleshell'. It was one of these central interlocking bones that I had found.

The shell of Sepiella immanis, a cuttlefish, is more complex than you would think at first.

We have now dwelt long enough in this less than salubrious, though highly productive habitat, and move onto an area with a more stable substrate, where we will stand less chance of sinking in.

An enigma explained: this is one of the plates that form the shell of a turtle.

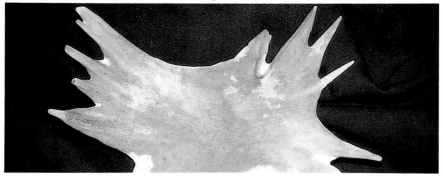

7 Fasht

The geological term 'fasht' is derived from an Arabic name for the limestone and coral aggregate that is found in a few locations in the southern Gulf and more frequently in Qatar, Bahrain and Saudi Arabia. It is perhaps a more common occurrence along the shores of the Red Sea, which is out of our sphere of coverage.

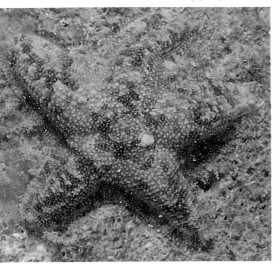

Small sea stars are found under fasht along with many other marine creatures.

At first glance it looks like crazy paving but when you lift up a piece of it at the edge of the shoreline, there is a real surprise in store. Underneath you'll see pieces of dead coral and living sponge which help to bind it together, and there are usually some of the larger purple sea squirts. Almost unnoticed are many more tiny creatures worthy of note. You may be lucky enough to find a small sea star, delicately patterned with green or red dots and blotches. There may also be a number of other sea stars and one with thin cylindrical arms called 'linckia' is easy to identify. Very often, brittle stars live under the fasht but with their long spindly arms they are difficult to handle and parts of them easily break off.

It looks like crazy paving and is unique to the Arabian coastline — fasht as seen here at RAK in the UAE.

With its long arms, the brittle star can move quickly over the surface of fasht.

Another of the sea stars to be found under fasht — Linckia.

Venus clams like this Callista lilacina are found in association with fasht.

One of the mistakes many people make is not looking long enough at the rock. It is all too easy to replace it without discovering some of the more unusual inhabitants. One of them is the chiton (pronounced 'kiton'). These tiny and primitive molluscs are not very active but if you look closely you may spot one clinging tightly to the underside of the fasht. They have eight overlapping plates surrounded by a girdle of flesh that may resemble snake skin or tufted bristles. There are several members of this class of mollusc in the Gulf, but they are notoriously difficult to identify.

Lepidozon luzonica is one of several chitons that can be found in this type of habitat.

Bivalve shells are often found hanging onto the fasht by means of their threadlike byssus. The ark shells are the easiest ones to find because of their shape. They are called by this common name because they resemble an ancient sailing boat, having a flat keel and gently curved edges.

Small tufts of what looks like a miniature coral is in fact a form of bryazoa. It can grow into a shape like a tiny tree and sometimes has a flat colourful covering which gives rise to one of its common names, that of 'sea-mat'. Where you find bryazoa you will sometimes discover a beautiful red sea slug that seems to feed on it.

So what may have looked like a lifeless piece of crumbly rock, in fact hosts a large number of tiny animals.

The spat of pearl shells start out in shallow zones such as the fasht belt, and eventually migrate into deeper water where they live for over a hundred years if left undisturbed.

8 Breakwaters and Jetties

Whereas fasht is a naturally occurring geological phenomena, breakwaters, harbour walls and jetties are purely man-made structures. None the less, they too hold a wealth of life.

Care should be taken when venturing out along the walls as the water swirls around the rocks and undercurrents can be dangerous. One of the first sights that greets you is the mass of algae that hangs down like green hair. It grows incredibly quickly and provides the first link in an ecological chain.

An edible limpet, Cellana radiata, is just one creature that feeds on the algae. It occurs on the structures in the southern Gulf, though it does not seem to be able to tolerate the higher salinity found north of Abu Dhabi.

Commonly found is a large green snail, Monodonta vermiculata, which also seems to have a range as far as Abu Dhabi. It is found around the Qatar Peninsula, though the specimens there seem to be much smaller and not as vividly marked as elsewhere, perhaps an effect of the high salinity recorded near Qatar.

One of the first molluscs to appear on any artificial habitat, Cellana radiata, is collected for food in parts of Asia.

Within days any rocks that are used as breakwaters will be covered with green alga.

Matuta is a shore crab that buries itself in the sand and has sharp spines at the edge of its carapace.

Turban shells, named for their resemblance to a turban, are another gastropod inhabitant of this zone which can easily be distinguished by the presence of a thick calcareous operculum. Perhaps this solid trap-door prevents them from being attacked by marauding crabs such as the Sally Lightfoot, a swift prancing shore crab that moves with deceptive speed around the rocks and will pick at just about anything.

Obviously the type of life to be found on a breakwater will depend on how long the breakwater has been established. On those that can be dated it would seem that life begins to colonise the rocks fairly quickly. Green weed starts to appear within two to three days, and this is followed soon after by other fast-growing species such as worms, barnacles and primitive molluscs. At a later date, more slow-growing animals start to move in

Trochus erythraeus, was named after the Red Sea, but is also very common in the Gulf.

and may even oust the early colonisers. This is something that seems to happen when pearl-bearing shells start to grow on underwater wrecks. The wing oyster, Pteria marmorata, is first on the scene, then it seems to get pushed out by the slower-growing radiating pearl shell, Pinctada radiata.

Euchelus asper, another top shell, was found to be abundant around the coast of Qatar.

On the rocks we do not get the wing oyster but the radiating pearl shell is frequently found. The alga that settles and covers the rocks so quickly is soon a magnet for other types of shore life. The large chitons, Acanthopleura vaillantii and Chiton peregrinus, feed on this green carpet at night and, like incessant lawn mowers, they remove much of it and keep everything neat and trim. Limpets, top shells, turban shells, planaxis shells, nerites and winkles are just some of the molluscs that create additional living space by clearing areas of fast colonising weed, thereby allowing other life forms the chance to move in. One of the later inhabitants, that does not make its presence known for at least a year, is also one that is much prized by shell collectors and beachcombers — the cowry.

This seems to be the case for many of the shells that are found in the Gulf, and it is interesting to form a comparison with the species found here and those in the Mediterranean. Long back in the mists of time, both the Red Sea and the Gulf area were fed by the waters of the Mediterranean when the region was part of a sunken basin. As the two areas parted company, many corresponding species grew in relative isolation and developed slightly different characteristics. When the southern Gulf opened its doors to the Indian Ocean an influx of new species came in to enrich the existing fauna and so make the Gulf one of the most prolific areas for marine life in general and molluscs in particular, for research workers to study in.

The large chiton, Acanthopleura vaillantii, grazes on the alga-covered rocks of breakwaters.

If breakwaters and harbour walls can provide entertainment for the collector, how much more is to be found on a naturally occurring rocky reef.

An attractive example of Chama pacifica, taken from under a breakwater rock.

Larger cowries, such as this Cypraea arabica or the Arabian cowry, do not appear on new breakwater sites for at least two years.

9 Rocky Beaches

Here then is the zenith for beachcombers because the number of species that can be found underneath exposed rocks far exceeds those present in any of the other zones so far discussed.

As they are the oldest and most stable of the habitats, they provide a home for a varied number of inter-dependent and inter-related groups of animals. What should not be overlooked is the fact that both the top and the bottom of the rock provide shelter for creatures large and small. You should also consider the area between the rocks as well as the sand underneath. Rocky reefs do vary a little, depending on the geological composition, but by and large there is a great similarity in the type of animal that likes to live there. In Qatar there is a particularly fascinating reef, especially when there is a low tide at night. The reef really does seem much more active then. This Fuwairat beach has yielded many shells — mostly of a small variety — crabs, flatworms, cucumbers, urchins, starfish, shrimps, sea slugs, even tiny fish. There in a single walk lasting two hours, I recorded no less than 82 species of mollusc alone!

Acorn barnacles grow over the rocks and are rarely covered with sea water.

Carpet anemones grow between the rocks and some have a commensal shrimp, Periclemenes brevicarpalis, living on them. This one is the larger and more attractive female.

A natural rocky beach near Sur in Oman.

Let us begin at the top of the rocks where there are several creatures to be found. The large volcano-shaped structures are the giant barnacle Tetraclita and there are some much smaller barnacles that may be found on the same rock. These are probably acorn barnacles, though as so often happens with common names, that term can also be used for the giant barnacle, leading to confusion and hence the use of Latin names for clear identification.

The jagged and razor-sharp edges of the rock oyster will be sure to get your attention if you inadvertently rub your hand over one. For this reason it is a good idea to wear gloves when moving the larger boulders about. Small white tubes that resemble a post horn are much in evidence and they are made by a small worm. Limpets can be found here, as can the giant chiton, Acanthopleura vaillantii, one of the species that was misnamed in the Gulf until recently. It was formerly called Acanthopleura haddoni by many authors and you may find it under that name in some of the publications on the area.

The powerful foot of a limpet allows it to have a watertight grip on rocks.

Flatworms ripple over the rocks and can be distinguished from nudibranchs by the absence of horns (rhinophores) and of the exposed gill circlet.

This stunning nudibranch has only been recorded in April and probably comes inshore to breed.

Shallow water near rocky beaches is home to many nudibranchs or sea slugs. This one is an aeolid.

Many brightly coloured sea stars live under or on rocks along the Musandam coast.

One of the dangers that lurk beneath exposed rocks is the fireworm which can inflict a painful sting with the bristly hairs along its edges.

The iridescence of this ragworm shimmers in the July sunlight.

All cone shells have a sting to stun their prey but, luckily, only a few are dangerous to man. Conus tesselatus, shown here, is harmless.

Cypraea ocellata occurs only outside the Gulf along the east coast of the UAE and into Oman.

Rocky Beaches

Lifting over some of the more manageable rocks is a somewhat laborious process but no matter how tiring it is you should always try to replace the rock exactly as it was found. Failure to do this will lead to natural predators moving in, and quickly finishing off whatever lives underneath. If they do not, the sun will. Under the rock there are a few dangers to be wary of. The first is the spiny sea urchin which is often encountered. It is not as prickly as the long-spined urchin, Diadema, that lives in shallow water and results in painful wounds if stepped upon, but it is still one to be avoided. Both these creatures belong to the same phylum and class as the sand dollars.

By far the most uncomfortable experience is given by the fireworm which is quite frequently seen under rocks along the east coast of the UAE and may well be present in other parts of the Gulf. Tiny hair-like bristles along the sides of the worm can detach easily and penetrate the skin even if you are wearing gloves. They can best be described as being like slivers of fibreglass and can cause a reaction if they get into the joints of the fingers. Normally the swelling will go away after a couple of days but sometimes lasts considerably longer. Apart from this nasty creature there is little else to worry about though you may find a couple of cone shells that are able to cause pain if they shoot their harpoon-like dart into you.

This fireworm, though very tiny, causes pain to the unwary.

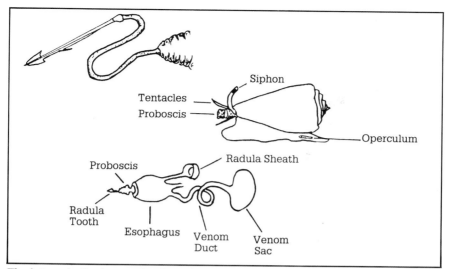

The internal stinging mechanism of a cone.

There are several species of cone present in the Gulf, all of which are able to kill their prey by the use of neurotoxin. This efficient poison is contained in a barbed hollow tooth that can be detached and fired off on the end of a thin cord of tissue. It is generally thought that only the larger cones pose a threat to humans and then only in a last desperate act of defence if removed from their habitat. But why take a chance? Far better to collect a dead specimen in good condition as these are quite adequate for building up a representative collection.

The cones so far found to be common under exposed rocks along the Oman and UAE coastlines include: Conus textile, C. pennaceus, C. striatus (all quite dangerous), C. coronatus, C. taeniatus, C. generalis, C. namoconus and C. flavidus. In the Gulf itself, C. episcopatus and C. textile can also be found and there are many more species that can occasionally be picked up along the beaches, some of them quite rare. Cones of course are not the only shells to be found — there are cowries, margin shells, mitres, nerite snails, ark shells and so on, all living under or on the rocks.

The textile cone, Conus textile, which causes intense pain, should be avoided.

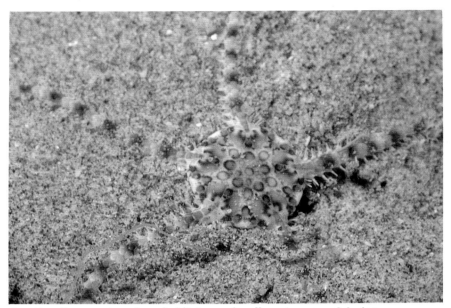

An attractive sea star whose habitat occasionally gets exposed during extreme low tides.

This huge sea star was found near Dibba.

A tiny crab bravely defends its territory on a rocky beach.

Decorator crabs adorn themselves with all sorts of material. One was even found with pieces of soft coral and black plastic.

An unidentified worm found under rocks on a Dubai beach.

A tube worm traps tiny particles in its fine fan-like structure.

It is difficult to get close to fan worms such as this Sabellastarte before they retract into their tubes.

Delicate Pharaonis australis — worms that are parasitic on tube anemones.

Rocky Beaches

In addition to these you may find several animals that seem strange when you first look at them. Flatworms, for instance, are one of the most colourful creatures in the sea and just seem to ripple across the rock in an attempt to find shade and protection. Most marine life has this desire for damp, dark places but there is one exception worth mentioning even though it does not occur in our area.

A green flatworm from northern France (Convoluta roscofensis) has found the answer to self-sufficiency. Its green colour is due to the presence of chlorophyll in the alga that it ingests. The structure of the cell walls of the alga changes after ingestion by the flatworm and the alga is not broken down in the normal way but remains in the body of the worm. The alga has now metamorphosed and cannot live outside the host as it has so radically changed from its original form. It goes on living and when the worm comes near to the surface of the water and into the sunlight the process of photosynthesis takes place inside its body. Masses of the species congregating near the surface of sandy pools in the warm sunshine give the appearance of seaweed which hence gain the worms a measure of protection through this unlikely camouflage. It is also probable that the

Flatworms have an incredible variety of colours.

A sponge crab desperately trying to conceal itself. Note also the brittle star on the sponge.

The body of this brittle star looks like an old man's face.

worm derives energy from the alga as it does not appear to need any other food source. As a result both the host and the alga live in a strange relationship that is called mutual symbiosis.

It may seem odd that the colourful and strange flatworms, which some consider to rival the beautiful sea slugs, should include in their number not only the self-sufficient green flatworm but the parasitic flukes and tapeworms. Other invertebrate animals (literally, without a backbone) that are present under the rocks are shrimps and crabs, anemones, sea squirts, sea cucumbers, brittle stars and even corals.

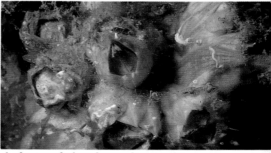
A cluster of giant barnacles, partly obscured by the yellow colonial sea squirts.

It is a world that you can get lost in for hours and before we move on there is an area around the rocks that is also worth looking at. Here we very often see clusters of green anemones. Their circlets of green tentacles are just below the surface of the water and look like tiny fairy rings contrasting with the sand. Occasional small fish wriggle around in the pools and the odd sand dollar may be found.

Under any well-worn piece of driftwood another surprise may be in store, in the form of many tiny, white structures with black hairy legs waving about. These are goose barnacles and they often attach themselves to floating debris and drift along in the currents until they finally get washed ashore.

10. Preserving your Collection

If you wish to collect any of the seaweed which forms the major part of the beachcrest, you will have to get up early in the morning as it soon gets dried in the hot sun. But what can we do with it when we do collect some that has been freshly thrown up by the waves? If you were in Japan you could use it for cooking, perhaps; in Scotland it is harvested for use as a fertiliser. There is a way of preserving it other than as a memory or with an expensive camera, but it is a simpler operation to perform than to describe! You need reasonable quality card or paper. You also need some thick absorbent material, such as paper towelling or medical gauze, and a large shallow tray. Now the only thing left is a specimen of fresh seaweed. Any of the three colour forms that can be found i.e. green, brown or red should retain some colour even after preservation. Once you have chosen your piece, take it home and carefully float it onto the card, which should be submerged in a shallow, water-filled tray. You then gently arrange your sample on the card until satisfied that you have it presented in its most attractive form. Now the trick is to remove the card gingerly, with the specimen in place on it, allowing as much of the water as possible to drain off. If you use a siphon to drain the water off, before removing the card and sea grass, you are more likely to achieve a perfect example.

Removing the card and siphoning off the water.

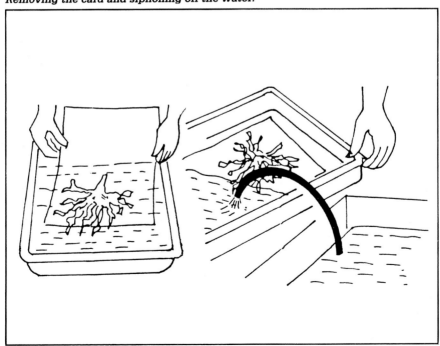

Placing an absorbent wad over the seaweed.

Once the weed has been drained and removed, the absorbent wad can be used to gently dab away any excess water that remains on the card. Next, place a fresh wad over the top of the weed, and then press it under a pile of newspapers or heavy weight. After every 12 hours, change the wad of absorbent material and very soon — usually after two to three days — you will have a fine dried and pressed specimen of seaweed. As always, try to record where and when you found the sample. There are many unusual and little-known species of seaweed here in the Gulf, and if you decide to make a collection of your specimens they can be presented in a neat book form. Put each of the cards into a clear plastic pocket and file it in a ring binder along with all the others.

Some of the common species from the Gulf that you can readily identify are peacock's tail, sargassum weed and oyster thief (so called because it develops on oyster beds, then some time later the weed becomes full of air and floats off, taking the oyster with it).

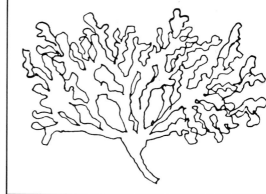

A dried and pressed specimen.

Looking like lettuce leaves, this species of sea grass was found in Hamriyah near Dubai.

Padina or peacock's tail weed is found throughout the Gulf.

The oyster thief, a highly specialised form of sea grass.

When viewed through a microscope it will be seen that sea grasses have thousands of tiny shells attached.

Tiny button tops require little preparation for the collection.

You will need a lot of patience to clean the giant murex shell, Chicoreus ramosus.

The inside of the shell of this Haliotis mariae, or abalone, is composed of nacre or mother-of-pearl.

It is almost impossible to remove the animal from the shell of the Babylon auger, or Terebra babylonia, so it is best to collect a dead sample.

Preserving your Collection

Apart from early morning hunts for seaweed, other collecting can be done at any time during the day. The drawback is that after a long hot afternoon on the beach, when you return home with a bag full of finds, the last thing on your mind may be a tedious cleaning operation. However, if you leave your specimens for too long they could be spoilt. All the effort expended collecting them would then be wasted. The best plan is to separate the various items and deal with those that will 'go off' — any freshly dead samples or ones with decaying matter in them; for example, part of an animal left inside a shell will cause a very powerful and unpleasant smell. I will not go into the preservation of live marine animals as this will only be of use to the very serious student. Dead shells can be left for a while but you should check them carefully as very often what appears to be an empty shell is, in fact, inhabited by a shy hermit crab. These crabs can suddenly become quite active and will scurry away and hide under any form of shelter. If you

It is worth spending time to clean the shell of the thorny oyster; the result will give it pride of place in your collection.

This collection was photographed in the Seychelles. Before you are tempted to buy, remember many items may not be allowed through customs there or at home, particularly tortoise shell ornaments.

Screw shells are aptly named. This one is Turritella maculata and comes from Bahrain.

leave them on the floor to be cleaned and the crab hides under your furniture it will eventually die, and the decomposing matter will make itself known in a most odorous way.

All that is generally needed is a thorough wash-off and, after drying, a light coating of baby oil. The use of varnish is frowned upon by serious malacologists but will certainly enhance the look of beach specimens in a bookcase.

If you have collected the old carapace of a crab or maybe the shell of a colourful crayfish, it only needs a gentle wash under the tap to remove sand and debris before it is ready to display. Again, the addition of varnish may enhance the crab shell or even those beach-worn shells in poor condition, but it should be remembered that in time the varnish will destroy the natural underlying lustre.

The cabinet used to house collections can vary according to the wishes and affluence of the individual. I have seen the most exquisite glass-fronted cabinets with subdued lighting to enhance the subtle colours of the specimens they contained. I have also seen variously collected items neatly labelled and kept in matchboxes and odd cartons which were of little cost, yet served the purpose quite adequately. One of the most unusual collection cabinets I ever came across was a replica Korean medicine chest with about 100 compartments, including some which were beautifully concealed to hide rarities in. It certainly was a conversation piece in itself. Perhaps the most practical containers of all are small perspex boxes in which the specimens can be seen yet which will protect them from dust and damage, the former a real problem in the Gulf. I have also used, for tiny micro specimens (i.e. shells and marine animals that measure less than 10mm in length), the plastic medicine capsules which can be obtained from some pharmacies. Inevitably your collection will outgrow the space that was originally allocated to it. Perhaps at this point it may be wise to limit yourself to a more narrow field of study rather than try to cover the whole spectrum of marine life.

Preserving your Collection

One area of collecting that requires a limited amount of space but is none the less very rewarding, is that of micro shells. It is quite a simple task to collect a box of damp sand from the area between high and low tide. If you then take this home and label the box with the date and the collection site it came from, you can deal with it at a later date. Once you find time to look at the sand, use a hand-held magnifying glass and a pair of tweezers. It will be easy to pick out the small shells which often have exquisite patterns. Of course, if you wish to appreciate their true beauty, a low-powered microscope will be a definite advantage. You can keep the micros in medicine capsules and, with the addition of a small collecting slip, you are on your way to starting a worthwhile shell collection. The only problem is that many of your finds will be difficult to identify with any certainty, as there is precious little literature on these fascinating and often delicate creatures.

The shell of the Turbo radiatus makes a particularly attractive addition to any beachcomber's collection.

Some micro shells live on other animals as parasites, such as Niso venosa which attaches itself to an astropecten sea star.

PERSONAL RECORD SHEET
NAME:

Common Name	Where Found	Date	Notes of Interest

11 Pollution

During the troubles which occurred in the northern Gulf, there was worldwide condemnation of the irresponsible actions that led to the widespread pollution of Gulf waters. A visible reminder of the now almost-forgotten oil spills continues to make its appearance on our beaches in the form of tar balls. It is often said that time is a great healer and apart from these unwanted visitations, it does seem to be. Crude oil in the form that was released into the Gulf had an immediate, damaging effect on marine life both above and below the surface. It floated on the surface for a very short period and unfortunately that was about the length of time it stayed in people's minds. Due to a complex action by the heat and a very fast growing alga, the slicks hardened and lost their bouyancy. The viscous cakes then sank and settled on the bottom of the sea bed. That was, however, not the end of the story as many may have thought. The action of the waves caused sand and shelly grit to adhere to these oil cakes as they lay on the seabed and they were eventually ground up into smaller and smaller pieces. It is these remnants which are still finding their way to the shoreline as tar balls and flecks. There is little we can do about this form of pollution other than to hope that it never happens again. In any event, the use of chemicals to disperse oil should not be considered a wise option as it often leads to greater cleaning-up costs and more damaging long-term effects on the environment.

Fishermen trolling for Bonito from a dhow, testament to the clean waters of the Musandam.

Nature has been able to deal with one of the most potentially catastrophic tragedies of this century with a speed not foreseen by most naturalists. This was helped by the relatively high temperatures, both ambient and seawater. It was also assisted by the extremely high salt content in the waters of the northern Gulf. However, we should not become complacent. In the event of any spill the offshore fisheries can become heavily polluted very quickly. It is the nature of bottom-feeding fish, such as the snapper, hamour, Sultan Ibrahimi, rabib and many of the rays, to take in mouthfuls of sand, sifting the edible from the non-edible matter. The waste is then ejected through their gills. Obviously, with the sand will be ingested particles of oil so it is imperative that long-term monitoring of catches is maintained.

Pollution

What then of the minor pollution problems that are within our ability to try and reduce, if not eradicate? There are many and it would be impossible to list them all so I will concentrate on the most obvious ones.

One of the products of our modern society that nature has not yet been able to deal with is the lasting effect of plastics. How many times have you seen supermarket bags being blown across the beach and ending up in the sea? These cheap plastic bags are a killer in the sea when they are mistaken by turtles for the large blue jellyfish that forms part of this much-loved sea creature's diet. I would like to see more supermarkets using paper sacks, as they are less environmentally damaging. Other plastics litter prime beach sites and with a little forethought they could be minimised. Larger containers, such as oil drums, are cheap and though not particularly attractive, they are practical as litter bins. Would it not be an ideal advertising opportunity for local traders if they were to provide a few with the name of their company emblazoned on them? Campaigns are undertaken by regional environmental committees in local schools and these should be encouraged in every way possible, perhaps with sponsorship of posters etc.

Turtles pick up blue plastic bags, mistaking them for jellyfish, and die as a result.

The coastal ecology of the Gulf is of a fragile nature.

It would also be an opportune time to conduct a survey of the remaining stretches of pristine beaches and declare the most important as marine sanctuaries, free from the damage caused by four-wheel-drive vehicles. Once the fragile surface crust is broken, immediate erosion begins. This will cause instability for small and salt-tolerant plants which attract insects and also bird life to our shores. Needless to say the grazing of animals quickly reduces any of the once abundant shrubs to nothing more than stunted twigs, and some strong fences are needed to keep out both wild and domesticated animals.

If this book helps you to know a little more about the things you can find when wandering along the beaches of the Gulf and its adjacent areas, then that will be reward enough for me. If it leads you to ask more questions than can be answered here, perhaps the short bibliography will point you in the right direction. And if by increasing your appreciation of the environment, it encourages you to preserve it, that will be an added bonus. Before we leave the shores of the Arabian Peninsula it may be timely to suggest a code of conduct that will help others to enjoy the shore-life as much as you do:

It is easy to wipe out unseen creatures, like this sea horse, by avoidable actions and then claim innocence.

Pollution

DO NOT OVERCOLLECT AND DO NOT TAKE MORE THAN YOU CAN HANDLE.

ALWAYS REPLACE ROCKS AFTER LOOKING UNDERNEATH THEM.

NEVER COLLECT ANIMALS LAYING EGGS.

PROTECT YOUR SPECIMENS FROM HEAT AND DAMAGE.

DO NOT COLLECT LIVE CONES OR DANGEROUS ANIMALS.

DO NOT PICK UP SEA SNAKES EVEN IF YOU ARE SURE THEY ARE DEAD — THE CHANCES ARE THEY ARE NOT!

TRY TO VARY YOUR COLLECTING SITES TO GIVE A BALANCED PERSPECTIVE.

ALWAYS RECORD YOUR FINDS AS SOON AS POSSIBLE.

CHECK THE TIDE TIMES.

MAKE SURE YOU PROTECT YOURSELF FROM THE SUN.

The Pericles tanker, ablaze in Gulf waters.

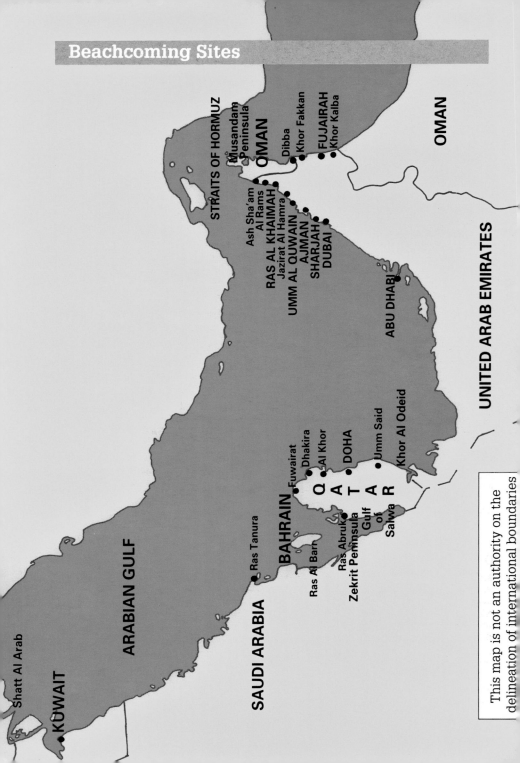

Reference Material

BIBLIOGRAPHY

Audubon Society: *Field Guide to North American Seashore Creatures* ISBN 0-394-51993-0

Barnes, Brian: *Coast and Shore* ISBN 0 946284-86-5

Bosch, Donald and Eloise: *Seashells of Southern Arabia* ISBN 1 873544 08 1

Cambell, A C: *The Seashore and Shallow Seas of Britain and Europe* ISBN 0-600-34396-0

Dawson, Yale E: *How to Know the Seaweeds*

Dipper, Frances and Woodward, Tony: *The Living Seas* ISBN 1 873544 10 3

Erwin, David and Piction, Bernard: *Guide to Inshore Marine Life* ISBN 0-907151 34-5

Friese, Eric U: *Sea Anemones*

Green, F and Keech, R: *The Coral Seas of Muscat* ISBN 0-946510 28-8

Lindner, Gert: *Seashells of the World* ISBN 0-7137 0844 1

Sharabati, D: *Red Sea Shells* ISBN 0-7103-0130-0

Shirai, Dr Shohei: *Ecological Encyclopaedia of the Marine Animals of the Indo-Pacific, Vol 1* ISBN 4-88024-092-3 C1640

Thompson, Dr T E: *Nudibranchs* ISBN 0-87666-459-1

Tokai University Press: *Marine Invertebrates (Japanese text with Latin names)* ISBN 4-486-00951-7 C0645 P2575E

Walls, Jerry G (ed): *Encyclopedia of Marine Invertebrates* ISBN 0-87666-495-8 H-951

Waterman: *The Physiology of Crustacea*

Zoological Society of London, Symposia of the: *Echinoderm Biology*

CLUBS AND SOCIETIES

Hawaiian Malacological Society
PO Box 22130
Honolulu
Hawaii 96823-2130
USA
Publishes worldwide information for shell collectors and sends out a monthly magazine.

American Malacologists
PO Box 1192
Burlington
MA 01803
USA

Netherlands Malacological Society
c/o Zoological Museum
Malacology Department
PO Box 20125
1000 HC Amsterdam
Netherlands

Sanibel-Captiva Shell Club
PO Box 355
Sanibel Island
Florida 33957
USA

Boston Malacological Club
c/o Ed Nieburger
PO Box 3905
Andover
MA 01810
USA

DEALERS

Mal de Mer Enterprises
PO Box 482
West Hempstead
NY 11552
USA
Also a book specialist

House of 10,000 Shells
PO Box 1437
Cairns
Queensland
Australia 4870

Kalika K Perera
95-2 Subadhrarama Road
Nugegoda
Sri Lanka

Cristina C Dayrit
PO Box 3
UP Post Office
Diliman
Quezon City
Philippines 3004

Femorale
CxP 15259
Sao Paulo SP
Brazil 01540

Acknowledgements

I would like to thank all my many friends in the UAE and throughout the Middle East, in particular members of the Dubai Sports Diving Club and Dubai Natural History Group. Special thanks go to Dick Macartney; Gerhardt and Doris (now in Austria); Monica Wingrove; Dieter and Caroline Lehmann; Ian and Trish Maiden; Ann and Dave Greedy; Julia Roles for having the patience to finalise the work so thoroughly; and last but not least, my wife Diane for all the nights spent exiled in my computer room.

Finally, thanks to

 Jebel Ali Hotel

whose support made possible the publication of this book.

Author

Tony Woodward returned to the UK in 1990 after many years in the Middle East. As a member of several regional natural history societies and chairman of the Doha and Dubai diving clubs, he gained a unique insight into the marine life of the Gulf.

He lead several expeditions into relatively unknown areas of the Musandam during extensive scientific research, and discovered two new species of Chiton, one of which bears his name (Acanthochitona woodwardi). Recently he appeared on Sky Television and the BBC and writes regularly for the magazine *Tropical Fish Hobbyist* in New Jersey.

Index of Common Names

abalone	20, 77	freckled cowry	22	sand collar	30
acorn barnacle	56, 57	garden slug	20	sand dollar	28, 29, 62
anemone	33, 69	gerbil	11	sand flea	37
ant	11	ghost crab	32	sand hopper	37
Arabian cowry	55	giant barnacle	57, 69	scallop	23
argonaut	26	giant chiton	57	screw shell	5, 75
ark shell	50, 68	gold-ringed cowry	21	sea cucumber	40, 56, 69
auger shell	20	goose barnacle	69	sea grass	70, 72, 73
Babylon auger	77	green snail	52	sea hare	11, 20
barnacle	46, 53	green turtle	47	sea horse	11
basket whelk	44	grey barnacle	32	sea mat	50
bivalve	17, 19, 23, 41, 50	grub	11	sea slug	20, 50, 56, 59, 69
blue jellyfish	81	hammer oyster	5	sea snake	28, 31, 83
blue swimming crab	38, 43	hamour	80	sea squirt	48, 69
Bonito	80	hermit crab	32, 37, 38, 44, 76	sea star	48, 49, 59, 64, 78
brittle star	48, 49, 69	kingfisher	43	sea urchin	56, 62
bubble shell	5, 20	kitten cowry	22	seaweed	70, 71, 76
bulbous olive	40	limpet	20, 52, 54, 57	shore crab	53
butterfly	11	long-spined sea urchin	62	shrimp	56
button shell	5, 76	mangrove tree	10, 42	snail	20
carnelian cowry	21	margin shell	68	snake	11
carpet anemone	56	micro shell	78	snapper	80
cerith shell	38, 43	mitre	5, 7, 20, 68	spider	18
chiton	19, 25, 50, 54	mollusc	4, 6, 12, 18ff, 33, 39	spiny murex	36
cicada	11		46, 50, 52, 54, 56	spiral staircase shell	33
clam	23, 36	money cowry	21	sponge	7, 32, 48, 69
cluster wink	45	moon snail	20, 30, 31, 37	sponge crab	69
cockle	25	moth	11	squid	26
collar	30	mud creeper	6, 43	starfish	56
conch	20	murex	5, 37, 41, 76	Sultan Ibrahimi	80
cone	5, 20, 61, 63, 83	mussel	23	sundial shell	5
coral	48, 50, 69	nautilus shell	26	tapeworm	69
corrugated clam	5	necklace shell	30	textile cone	63
cowry	5, 20, 21, 54, 63	nerite	54	thorny oyster	5, 7, 41, 74
crab	11, 18, 22, 32, 38, 46, 56	nerite snail	68	thrush cowry	21
	65, 77	octopus	26	top shell	5, 6, 7, 54
crayfish	77	olive shell	39, 40	tree rat	11
curlew	11, 43	oyster	23, 46	tube anemone	67
cuttlefish	26, 27, 47	oyster thief	71, 73	tube worm	66
decorator crab	65	peacock's tail weed	34, 71, 72	turban shell	5, 53, 54
desert hyacinth	10	pearl oyster	6, 23, 32, 35	turrid shell	5, 7
diatom	25		51, 53, 54	turtle	81
dove shell	44	pink barnacle	32	tusk shell	5, 25
dragonfly	11	planaxis shell	46, 54	univalve	20
fabulous kitten cowry	22	plover	43	Venus comb murex	36
false winkle shell	45	precious wentletrap	33	wading bird	11, 43
fan worm	67	rabib	80	wasp	11
fiddler crab	43	ragworm	60	wentletrap	5, 7, 33
file shell	25	rays	80	window pane oyster	43
fireworm	60, 62	reef heron	43	wing oyster	53, 54
flatworm	56, 58, 68, 69	rock oyster	57	winkle	20, 46, 54
fluke	69	Ruschenbergs scallop	41	worm	11, 53, 57, 66, 67, 69
fluted clam	23	Sally Lightfoot crab	53	zigzag cowry	6

THE ARABIAN HERITAGE SERIES

Arabian Profiles
edited by Ian Fairservice and
Chuck Grieve

Bahrain – Island Heritage
Enchanting Oman
Land of the Emirates
by Shirley Kay

Kuwait – A New Beginning
by Gail Seery

Dubai – Gateway to the Gulf
edited by Ian Fairservice

**Abu Dhabi – Garden City of
the Gulf**
by Peter Hellyer and
Ian Fairservice

Fujairah – An Arabian Jewel
by Peter Hellyer

Portrait of Ras Al Khaimah
Sharjah – Heritage and Progress
by Shirley Kay

Gulf Landscapes
by Elizabeth Collas and
Andrew Taylor

Birds of the Southern Gulf
by Dave Robinson and
Adrian Chapman

**Falconry and Birds of Prey in
the Gulf**
by Dr. David Remple and
Christian Gross

The Living Desert
by Marycke Jongbloed

The Living Seas
by Frances Dipper and
Tony Woodward

Mammals of the Southern Gulf
by Christian Gross

Wings Over the Gulf
Seafarers of the Gulf
by Shirley Kay

**Architectural Heritage of
the Gulf**
by Shirley Kay and
Dariush Zandi

Sketchbook Arabia
by Margaret Henderson

Travelling the Sands
by Andrew Taylor

**Juha – Last of the
Errant Knights**
by Mustapha Kamal,
translated by Jack Briggs

Storm Command
Looking for Trouble
by General Sir Peter de la Billière

This Strange Eventful History
by Edward Henderson

**Zelzelah – A Woman Before
Her Time**
by Mariam Behnam

**The Oasis – Al Ain Memoirs of
'Doctora Latifa'**
by Gertrude Dyck

The Wink of the Mona Lisa
by Mohammad Al Murr,
translated by Jack Briggs

Fun in the Emirates
Fun in the Gulf
by Aisha Bowers and
Leslie P. Engelland

Premier Editions

A Day Above the Emirates
A Day Above Oman
by John Nowell

Forts of Oman
by Walter Dinteman

Land of the Emirates
by Shirley Kay

Seashells of Eastern Arabia
edited by S. Peter Dance

50 Great Curries of India
by Camellia Panjabi

Dubai – A Pictorial Tour

**The UAE –
Formative Years 1965-75**
by Ramesh Shukla

MOTIVATE
PUBLISHING

Arabian Heritage Guides

**Off-Road in the Emirates
Volumes 1 & 2**
by Dariush Zandi

Off-Road in Oman
by Heiner Klein and
Rebecca Brickson

**Snorkelling and
Diving in Oman**
by Rod Salm and
Robert Baldwin

**The Green Guide to
the Emirates**
by Marycke Jongbloed

**Beachcombers' Guide to
the Gulf**
by Tony Woodward

On Course in the Gulf
by Adrian Flaherty

Spoken Arabic – Step-by-Step
by John Kirkbright

The Thesiger Library

Arabian Sands
The Marsh Arabs
Desert, Marsh and Mountain
My Kenya Days
Visions of a Nomad
by Wilfred Thesiger

The Thesiger Collection
a catalogue of
photographs
by Wilfred Thesiger

Thesiger's Return
by Peter Clark
with photographs by
Wilfred Thesiger

Thesiger
by Michael Asher

Arabian Albums

Written and photographed
by Ronald Codrai:

Dubai
Abu Dhabi
**Sharjah and the North-East
Shaikhdoms**
Travels to Oman